Gemini Comics

I0481594

ISBN 978-1983781292

Manufactured in the United States of America

For entertainment purposes only

Uncle Newton

in
CONDUCTOR of the
UNIVERSE

Uncle Newton

Newton wanted to know
how the universe worked.
His insatiable curiosity led him
to ask questions
and perform many
experiments, all
leading him to a multitude of
very important
scientific discoveries.
His findings helped describe
the way things are
naturally configured –
giving humankind a deeper
understanding of nature, and
even setting the foundation
for the modern world.

Newton was
fascinated
with the universe,
and wanted
to know how
it worked.

He often
compared the
universe
to a clock -
brilliantly
configured
and *always*
in motion.

*Newton's discoveries also helped to describe the planets' orbits, and even helped prove that the sun is at the center of the solar system.

Newton was
very crafty –
he loved to build
and design
all kinds of unique
and interesting
contraptions.

Newton was born and
raised in England,
in a home called
Woolsthorpe Manor.
Some of his greatest
thoughts, discoveries
and achievements
were made there.

He secretly practiced alchemy, thinking he could transform ordinary substances into precious metals.

He wrote arguably THE most important book in all of science – *La Principia Mathematica*. It elegantly described how things move and interact with each other in the universe.

AUTHOR

For example,
he wrote
that *nothing*
in life gets done
without a PUSH
or a PULL, in
a concept called
inertia.

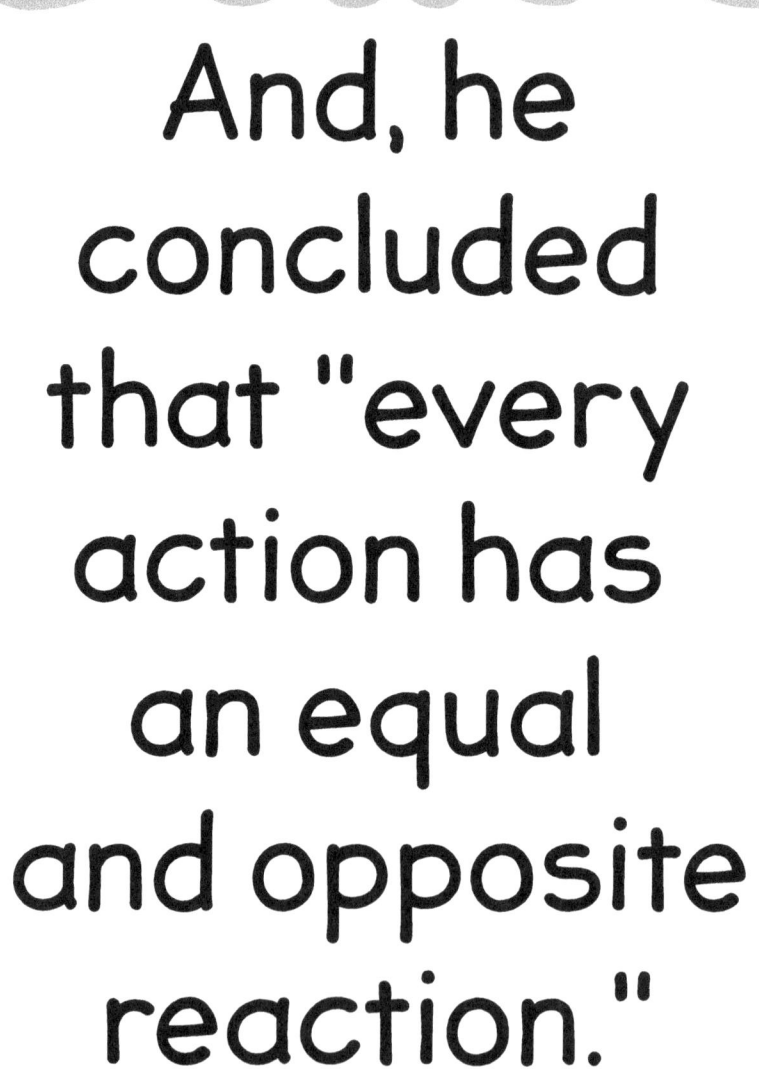

And, he concluded that "every action has an equal and opposite reaction."

After
experimenting
with a glass prism,
he discovered
that white
light is composed
of all the
colors of
the spectrum.

He developed a
new mathematical
system of tracking
and measuring
change (fluxions),
called "calculus."

He was able to peer DEEP into the COSMOS, thanks to his invention of the REFLECTING telescope.

He figured out
that everything in
the universe is
attracted to
each other by
a force called
"Gravity."

He passionately
taught mathematics
at Cambridge
University.

Later in life, Newton became the president of the Royal Society – an elite association that attracted the brightest scientific minds of its day. The organization was at the forefront of natural discovery and scientific research.

He also became
Master of the
Royal Mint,
improving the
efficiency and
effectiveness of
the currency
system in London.

Isaac Newton
received the greatest
honor by becoming
"SIR" Isaac Newton
when he was
knighted for
his lifetime
achievements and vast
contributions
to the world of
science.

Science was Newton's passion. But, an even deeper fascination was his connection to the divine – the entity that created the universe.

The End

www.ingramcontent.com/pod-product-compliance
Lightning Source LLC
Chambersburg PA
CBHW081622220526
45468CB00010B/2993